化石猎人在工作

Fossil Hunters at Work

[英] 露丝・欧文/著
刘颖/译

江苏凤凰美术出版社

全家阅读
小贴士

★ 每天空出大约10分钟来阅读。

★ 找个安静的地方坐下，集中注意力。关掉电视、音乐和手机。

★ 鼓励孩子们自己拿书和翻页。

★ 开始阅读前，先一起看看书里的图画，说说你们看到了什么。

★ 如果遇到不认识的单词，先问问孩子们首字母如何发音，再带着他们读完整句话。

★ 很多时候，通过首字母发音并听完整句话，孩子们就能猜出单词的意思。书里的图画也能起到提示的作用。

最重要的是，感受一起阅读的乐趣吧！

扫码听本书英文

Tips for Reading Together

- Set aside about 10 minutes each day for reading.
- Find a quiet place to sit with no distractions. Turn off the TV, music and screens.
- Encourage the child to hold the book and turn the pages.
- Before reading begins, look at the pictures together and talk about what you see.
- If the child gets stuck on a word, ask them what sound the first letter makes. Then, you read to the end of the sentence.
- Often by knowing the first sound and hearing the rest of the sentence, the child will be able to figure out the unknown word. Looking at the pictures can help, too.

Above all enjoy the time together and make reading fun!

目录 Contents

你想成为一名化石猎人吗？
化石猎人挖掘恐龙化石。

Would you like to be a **fossil** hunter?

Fossil hunters dig up dinosaur fossils.

化石猎人
a fossil hunter

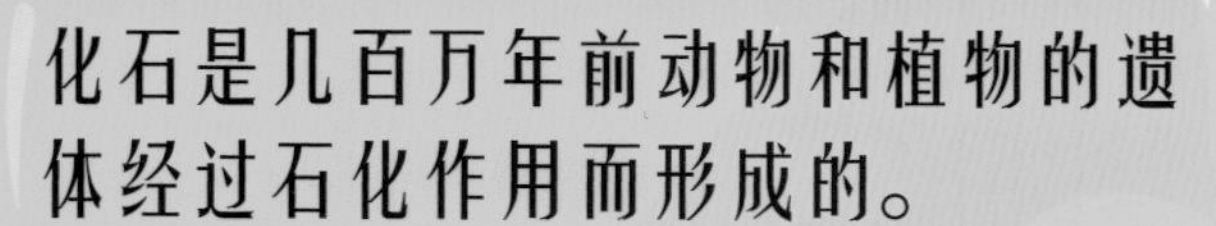

化石是几百万年前动物和植物的遗体经过石化作用而形成的。

Fossils are the rocky **remains** of animals or plants that lived millions of years ago.

这是哪种恐龙？
What Kind of Dinosaur?

科学家仔细观察化石的大小和形状。

Scientists look carefully at the size and shape of a fossil.

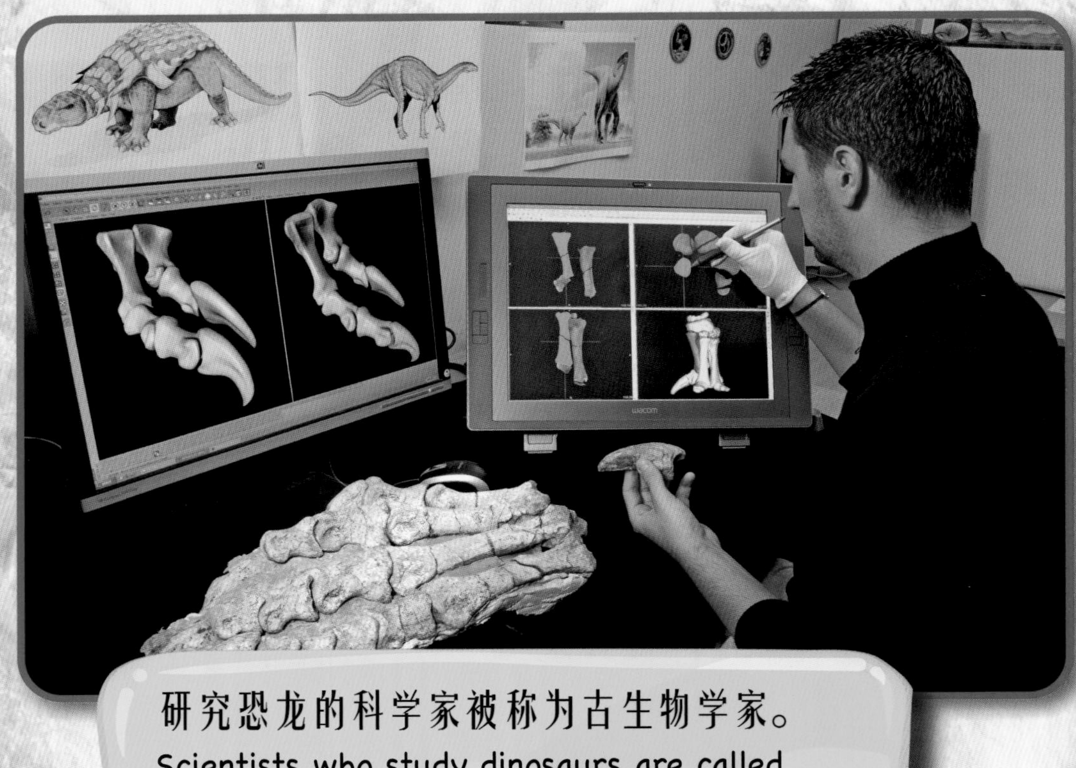

研究恐龙的科学家被称为古生物学家。

Scientists who study dinosaurs are called palaeontologists (pay-lee-un-TOL-uh-jists).

他们会判断这块化石属于哪种恐龙。

They work out what kind of dinosaur the fossil belonged to.

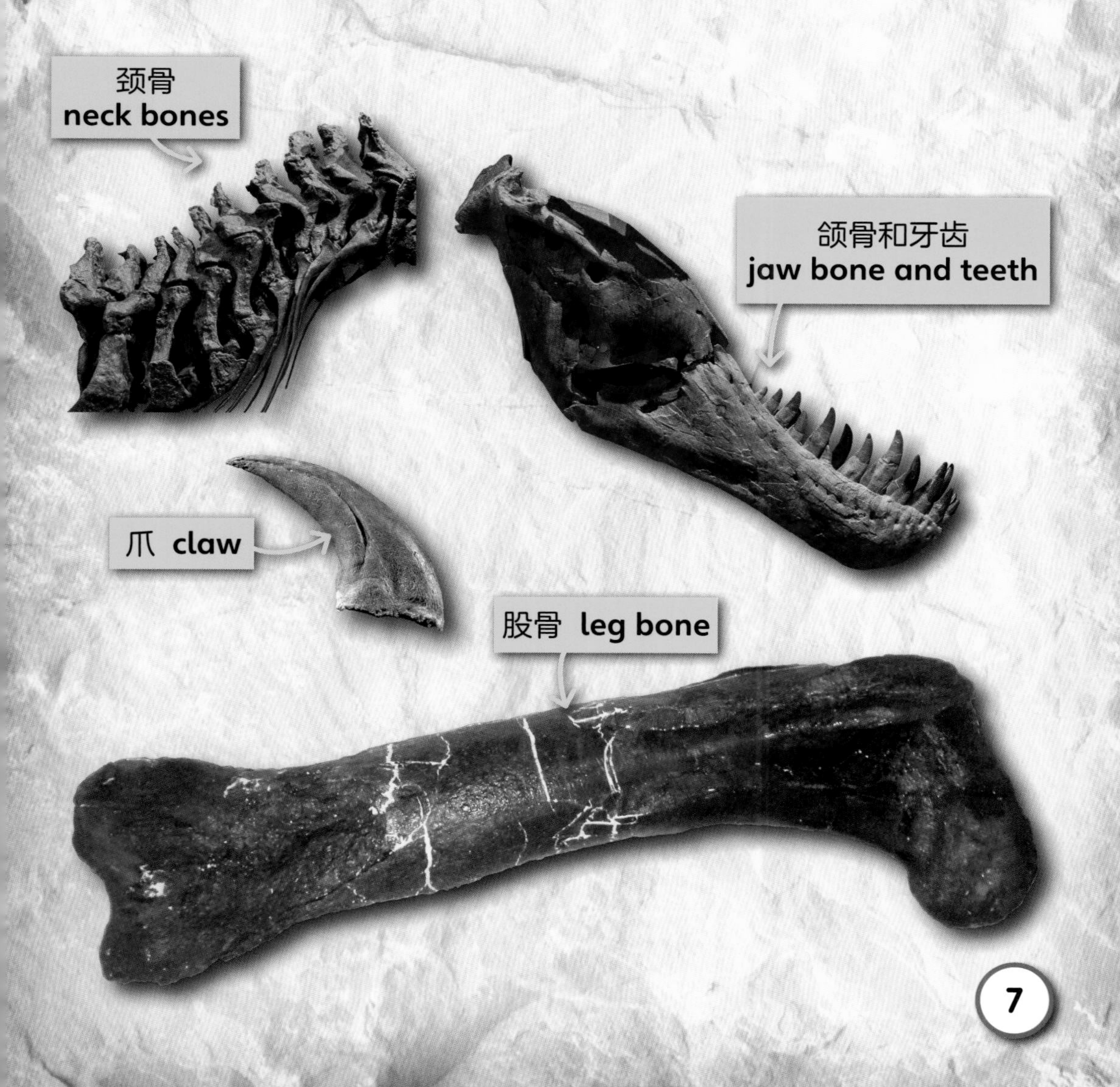

挖掘恐龙化石
Digging Up the Dinosaur

化石大多是在远离城镇的石质荒漠中被发现的。

Fossils are often found in rocky **deserts**, far from a town.

当科学家发现一块恐龙化石时，
他们将继续挖掘以寻找恐龙的其余部分。

When scientists find one fossil,
they dig to find the rest of the dinosaur.

挖掘恐龙化石的地方就是“挖掘现场”。
The place where a dinosaur is dug up is called the “dig site”.

开始挖掘
The Dig Begins

科学家使用斧镐、手提钻和铲子挖出岩石。

The scientists dig up the rock with axes, jackhammers and shovels.

科学家可能会用挖掘机从挖掘现场挖走成吨的岩石。

Scientists may use a digger to remove tons of rock from a dig site.

科学家有时会挖到霸王龙的头骨！

Sometimes the scientists find the skull of a Tyrannosaurus rex!

化石猎人在工作
Fossil Hunters at Work

接着，科学家移除化石周围的岩石。

Next, the scientists remove the rock close to the fossils.

他们使用小泥铲、凿子和刷子小心翼翼地移除岩石。

They do this carefully with small trowels, chisels and brushes.

科学家还要绘制挖掘现场的地图。
地图上标出了每块化石的发现地点。

Scientists draw a map of a dig site.

This shows where each fossil was found.

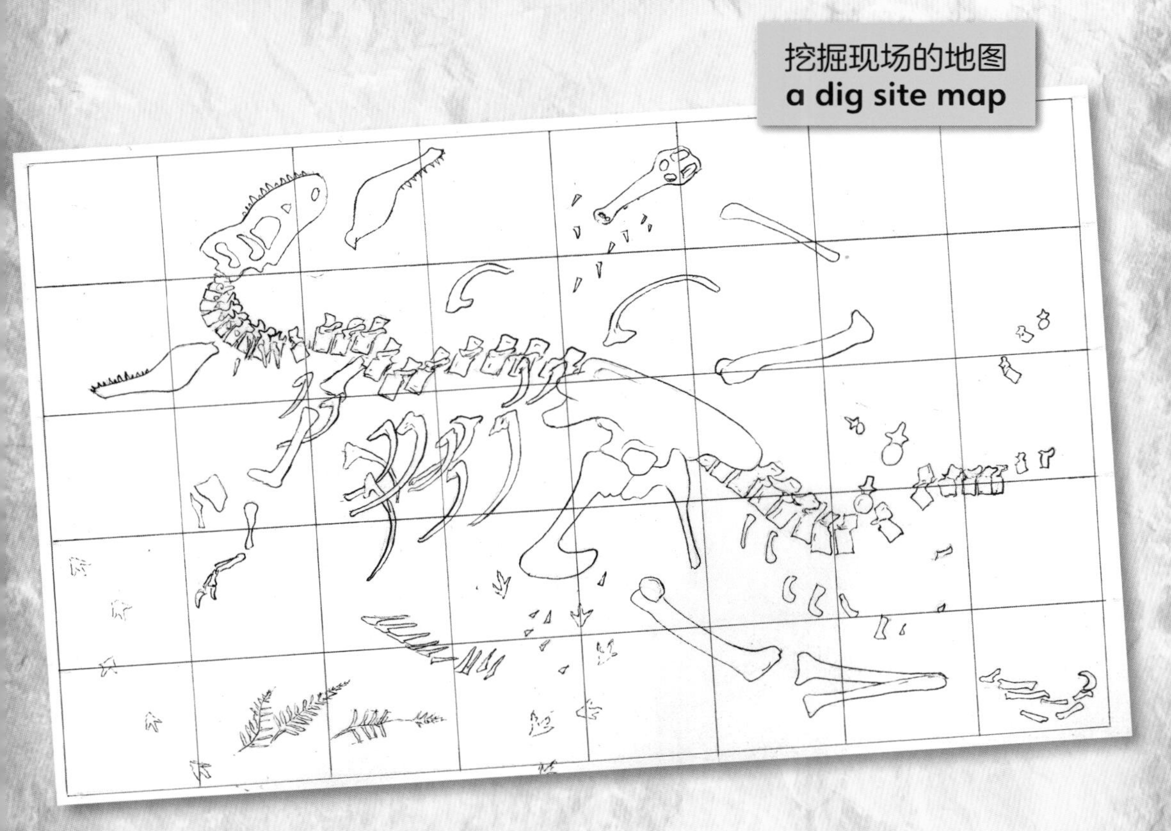

挖掘现场的地图
a dig site map

保护化石
Protecting a Fossil

一具霸王龙头骨从岩石里被挖出。
科学家用湿石膏将它包裹住，以提供保护。

A T. rex skull is dug up inside a lump of rock.
The scientists cover it in a wet **plaster** jacket to protect it.

石膏干了以后会变硬。

As the plaster dries, it gets hard.

被石膏外壳包裹的头骨
skull inside a plaster jacket

寻找线索

Looking for Clues

化石猎人寻找附近的其他化石。

Fossil hunters look for other fossils nearby.

他们有时会找到植物化石。
植物化石告诉他们数百万年前生长了哪些植物。

Sometimes they find fossil plants.
This tells them what kinds of plants grew millions of years ago.

如果他们找到了鳄鱼牙齿的化石，他们就会知道鳄鱼吃了恐龙的尸体。

If they find fossil crocodile teeth, they know that crocodiles ate the dead dinosaur's body.

送去博物馆
Off to the Museum

科学家用一台牵引机将霸王龙头骨吊到卡车上。

The scientists use a tractor to lift the T. rex skull onto a truck.

然后将它送去博物馆。

Then it is taken to a **museum**.

在博物馆里，科学家切除包裹化石的石膏。

At the museum, scientists cut away the plaster from the fossils.

会见霸王龙
Meet a T. Rex

接着，科学家将岩石炸开。
然后，他们使用牙科工具将碎石剔掉。

Then the scientists blast away the rock.
Next, they use a dental tool to pick away the rock.

牙科工具
dental tool

最后，化石在博物馆里展出，大家都可以参观霸王龙的头骨。

Then the fossil can go on show in the museum and people can see the T. rex skull.

词汇表 Glossary

荒漠 desert

干燥的陆地，植物罕见且降雨稀少。

Dry land with few plants where very little rain falls.

化石 fossil

存留在岩石中几百万年前的动物和植物的遗体。

The rocky remains of an animal or plant that lived millions of years ago.

博物馆 museum

研究和展出化石和艺术品等有趣物品的场所。

A building where interesting objects, such as fossils and art, are studied and displayed.

石膏　plaster

由石头、沙子和水形成的液态混合物，干了以后会变硬。

A liquid mix of rock, sand, and water that turns hard when it dries.

遗体　remains

尸体的全部或部分。

All or part of a dead body.

科学家　scientist

研究自然和世界的人。

A person who studies nature and the world.

恐龙小测验 Dinosaur Quiz

① 化石猎人的工作是什么？

What do fossil hunters do?

② 化石是什么？

What are fossils?

③ 研究恐龙的科学家被称为什么？

What are scientists who study dinosaurs called?

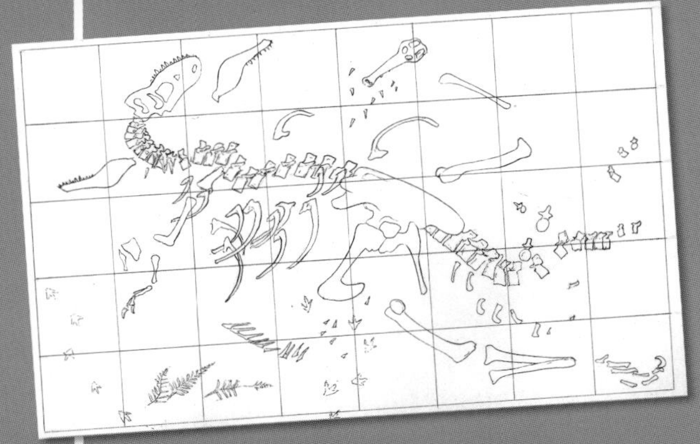

④ 科学家为什么要绘制挖掘现场的地图？

Why do scientists draw a map of the dig site?

⑤ 你想成为化石猎人吗？

Would you like to be a fossil hunter?